CONTENTS

Contents	1
Tinkerbell – The Beginning	2
The History of the Tinkerbell Steam Locomotive – Part 1	5
Developing the Tinkerbell Concept	17
The History of the Tinkerbell Steam Locomotive – Part 2	25
Driving & Firing 'Tinkerbell' – Peter Jackson	40
Memorable Days with 'Tinkerbell' – Peter Jackson	43
The History of the Tinkerbell Steam Locomotive – Part 3	45
Building a Tinkerbell	57
The History of the Tinkerbell Steam Locomotive – Part 4	61
Photo Index	67
Bibliography	68

All rights reserved. No part of this book may be reproduced or transmitted in any form or by any means, electronic or mechanical, including photocopying, recording or by any information storage and retrieval system, without permission from the Publisher in writing.

Copyright © 2018 Paul Ash

Published by Independent Publishing Network.

First Published 2018

Printed by Direct Offset, Glastonbury BA6 9DR [01458 831417]

ISBN: 978-1-78926-609-2

Author: Paul Ash. (Care of) The Moors Valley Railway, Moors Valley Country Park, Horton Road, Ashley Heath near Ringwood, BH24 2ET. [01425 471415]. Please direct all enquiries to the author.

Cost: £6.50 per issue

TINKERBELL – THE BEGINNING
(How it all came about)

The origins of what is known today as a Tinkerbell class steam locomotive date back to 1966 and to the builder of the original *Tinkerbell*, namely one Roger Marsh. At the time Roger had a small 5″ gauge 0-4-0WT but was inspired to build something much larger to run on a friend's railway. The loco chosen for the model was one of the 18″ gauge Beyer Peacock locomotives produced for the Lancashire & Yorkshire Railway in 1887 and used as the Horwich Works shunters. Roger decided this could be modelled half full size running on 7¼″ gauge track. These locos featured a marine type boiler, large dome and a tall chimney, all features to be seen on *Tinkerbell*.

18″ gauge Horwich Works shunter Wren *is pictured here as part of the collection at the National Railway Museum. This loco ran with a saddle tank in addition to the standard well tank.* (Photo: Paul Ash)

Construction was very much a case of using whatever materials were cheaply available, with Roger calling on friends and favours in many cases. The original wheels for example came from an old mobile crane, the water filler cap came from the scrap box but started life on a fire hydrant, and the smokebox door complete with its polished brass knob from a WW2 Motor Torpedo boat.

Roger's original concept was for the driver to sit in a tender running behind the loco and in fact the tender was constructed first. It was then Roger was introduced to the work of Sir Arthur Heywood and his minimum gauge railway from the late 1800's. Whilst Sir Arthur Heywood worked in 15˝ gauge, Roger saw the potential to scale this down and combined this influence into the work he had already completed. It was at this point the frames were extended and a rear pony truck added to make an 0-4-2T with the driver sitting on the loco. The Tinkerbell concept was born.

Left: Ursula *is a new build recreating one of the original Minimum Gauge locos constructed by Sir Arthur Heywood.* (Photo: Paul Ash)

Notable features taken from Heywood locos were large side tanks to hold the water and coal, a weatherboard style cab to protect the driver and Heywood style valve gear, again all features seen on the original *Tinkerbell*. Back to the build and by now the cylinders were fabricated by a friend and Roger had completed building the marine boiler.

The marine boiler on *Tinkerbell* features a cylindrical firebox behind the tubeplate. The grate, heat baffle, ashpan and fire door backplate are then inserted as one complete unit into the large cylindrical firebox. This type of boiler is cheap to produce compared to a locomotive type boiler.

Right: Tinkerbell *is still running with a marine type boiler today although it is no longer the original.* (Photo: Paul Ash)

Roger completed *Tinkerbell* in 1968 ready for the first steam tests. After construction Roger continued to develop his ideas and a number of different modifications were made such as fitting a superheater to the boiler, improved springing and changing the complex rear pony truck design. One feature that failed to be a success was a variable blast nozzle, which if used whilst the engine was running could result in the draft from the engine's exhaust being directed back into the cab with serious safety concerns.

Roger would not only run the new loco, named *Tinkerbell* after his pet dog, at his friend's line but also take the loco further afield to other railways and special events. This started to generate considerable interest in the new locomotive and pave the way for future new locomotive construction.

Left: *A young Roger Marsh pictured sitting in* Tinkerbell *shortly after construction.* (Photo: Unknown)

Tinkerbell was sold by Roger to Jim Haylock in 1978, and as we shall see Jim took the concept and continued to develop and promote the idea of a minimum gauge railway.

Today a Tinkerbell design is a term commonly used to refer to any 7¼" gauge locomotive of narrow gauge proportions where the driver sits inside the cab of the locomotive. The Moors Valley Railway in Dorset is now the home of *Tinkerbell*, alongside three other standard Tinkerbell design locos, and numerous other larger locos based on the original concept.

THE HISTORY OF THE TINKERBELL STEAM LOCOMOTIVE – PART 1

(A potted history lesson in four parts)

Tinkerbell herself, now under the ownership of Jim Haylock, was delivered to her new owner in 1978. At this time *Tinkerbell* also received a major overhaul including the fitting of a new boiler, fly cranks and a full repaint into a maroon livery.

Keith Watson driving a newly delivered Tinkerbell. (Photo: Jim Haylock)

Jim initially ran *Tinkerbell* at the Malden & District Society of Model Engineers railway at Thames Ditton in Surrey, along with visiting a number of 7¼" Gauge Society events, whilst at the same time helping to promote his new business, Narogauge Ltd, selling sit in covered coaches to run with the narrow gauge locos. At the same time Jim also purchased *Talos* from the first production batch of locomotives.

Right: Tinkerbell *is pictured here with the covered coaching stock built by Jim Haylock for sale under the name Narogauge Ltd.* (Photo: Russ Green)

In 1981 Jim took over the running of the miniature railway at Tucktonia in Christchurch, Dorset. Tucktonia was a model village and forerunner of the modern day theme park, which featured models of famous London landmarks such as Big Ben and Buckingham Palace. The line had been constructed in 1973 as 10¼″ gauge and this was re-laid by Jim to 7¼″ gauge before opening in April 1981. Whilst at Tucktonia *Tinkerbell* was the main stay of passenger hauling until such time as larger and more powerful locomotives were constructed to run on the line.

As built *Tinkerbell* was fitted with a weatherboard style cab, however a fully enclosed cab had also been made by Roger Marsh to a design by Jim Haylock, so *Tinkerbell* could run on wet days whilst providing a level of comfort for the driver.

It was in this guise (and appropriately in the pouring rain) that *Tinkerbell* managed to make a TV appearance on the BBC children's TV show Multi-Coloured Swap Shop, hosted by Noel Edmonds and Keith Chegwin. The line at Tucktonia closed in 1985.

Right: Tinkerbell *seen running with the fully enclosed cab.* (Photo: Paul Ash)

1986 saw Jim Haylock open a new railway at the Moors Valley Country Park in Dorset and *Tinkerbell* once again proved her worth in the construction of the new line.

Left: Tinkerbell *was regularly used in the construction of the Moors Valley Railway transporting ballast and other materials to site in a set of 4-wheel tipper wagons, a true narrow gauge spectacle.* (Photo: Russ Green)

Left: Tinkerbell *has also been seen running cab-less on those really hot summer days as seen here with Jim Haylock driving.* (Photo: Paul Ash)

2018 marks the 50th anniversary of the construction of the original *Tinkerbell* and the loco is receiving a significant major overhaul including boiler re-tube (whilst retaining her marine boiler), new wheel sets, cylinder re-bore with new liners and some replacement valve gear parts. A full repaint is also planned.

The aim is to complete the rebuild by the 50th Anniversary celebrations at the Moors Valley Railway in September 2018, where it is planned to bring together over 20 Tinkerbell class locos for this special event.

Tinkerbell *is undergoing a major rebuild in 2018. As you can see from this photo taken in July 2018 the rebuild is being completed to a very high standard however there still remains a lot of work to complete before* Tinkerbell *steams again.* (Photo: Paul Ash)

Following on from the interest generated by the newly built *Tinkerbell*, Roger Marsh started taking orders in 1975 for a batch of five production Tinkerbell class locomotives. These were eventually completed by Roger in 1978 at his Britannia Works before being shipped to their new owners later that year.

The basic specification was similar to the original *Tinkerbell* with an 0-4-2T wheel arrangement, 2½″ bore x 4¼″ stroke cylinders, wheel diameter 10¾″ (increased from 10″ on *Tinkerbell*) and spaced at 17″ fixed wheelbase. Heywood valve gear and a marine boiler completed the package. Whilst all five were built to the same basic specification there were some detail differences between the locomotives.

Roger Marsh's workshop in 1978. Left to Right: Susan Jane, Adam, Sir Robert *and* Gurkha. (Photo: Jim Haylock)

Right: *3 new boilers ready for fitting to* Gurkha, Adam *and* Sir Robert. (Photo: Jim Haylock)

8

TALOS

Talos was the first of the production locos to be delivered in 1978. It was painted in a lined red livery and was fitted with a full width front spectacle plate, initially featuring very small round windows for the driver to see out. These have subsequently been enlarged to give a better view and a cab roof added for those wet days to keep the driver dry.

Jim Haylock (left) *and Roger Marsh* (right) *in* Talos *and* Tinkerbell *respectively at the Echills Wood Railway, Stoneleigh, Warwickshire in 1978.* (Photo: Jim Haylock)

Left: Talos *suffered a loose crankpin during steam trials which required some improvisation by Roger to weld back up.* (Photo: Jim Haylock)

Initially running alongside *Tinkerbell* at Tucktonia, it was during this time *Talos* was converted from a marine boiler to a locomotive boiler by adding a locomotive firebox, the conversion taking place whilst keeping the original boiler barrel.

Talos *is pictured here running with a cab roof. The driver is Mike Haylock, brother to Jim Haylock.* (Photo: Paul Ash)

Talos (driven by Jim Haylock) also took a prominent role advertising Tinkerbell class locomotives in the Severn Lamb railway catalogue in the early 1980's. Severn Lamb from Stratford on Avon are probably more famous for building a pair of Big Boy steam locomotives, one of which ran on the Dobwalls Miniature Railway in Cornwall, sadly now closed and the loco exported to Australia.

(Photo: Severn Lamb Ltd)

Since moving to the Moors Valley Railway both *Tinkerbell* and *Talos* have continued to see regular use on gala days, double heading on passenger services and hauling freight trains. *Talos* however has gained a special place at the railway being the loco of choice for their driver experience days.

The driver experience gives participants the chance to drive *Talos* under instruction for the day, and then gain a certificate at the end. The course has proved particularly popular as birthday and retirement gifts, or those just wanting to experience the thrill of driving a real steam locomotive on a fully signalled railway. At least one participant has gone on to build two Tinkerbell locomotives themselves as a result of driving *Talos* on this course. Since receiving a major rebuild *Talos* continues to operate regularly at the Moors Valley Railway.

Talos *is pictured here after a day's work on the Driver Experience course at the Moors Valley Railway.* (Photo: Paul Ash)

SUSAN JANE

Susan Jane was turned out in what was to become the traditional Tinkerbell style with a marine boiler, Heywood valve gear and weatherboard cab. Livery was blue fully lined with a black border and the chimney was topped by the standard flared brass cap. The loco was purchased through Cromer White Railways by Lawrence Martin for use on his newly laid Wayside Light Railway in Kent. Purchase price at the time was £2,500 which equates to a value of £20,000 today.

The Wayside Light Railway is a private garden railway and was initially constructed as a simple line running from the front to the back garden. Today the railway has significantly expanded with trains running out and back from the main station at Wayside around a lengthy and complex route. The line features steep gradients, bridges, tunnels, two other stations and is of course fully signalled. *Susan Jane* is still a regular performer on the line some 40 years later.

Susan Jane has retained her original appearance throughout her life with very little changes. Today she is easily recognisable as the same loco from 40 years ago, albeit now with a cast chimney top, and can still be seen running at Wayside.

Left: Susan Jane *running on the Wayside Light Railway in 2015.* (Photo: Paul Ash)

Susan Jane *on a visit to the Moors Valley Railway.* (Photo: Paul Ash)

SIR ROBERT

Purchased at the same time as *Susan Jane*, and for use on the same Wayside Light Railway, *Sir Robert* has undergone significant change from the loco that was first delivered back in 1978.

Lawrence Martin takes delivery of Sir Robert *from Roland White of Cromar White Railways.* (Photo: Jim Haylock)

When new *Sir Robert* featured a half cab, was painted in Stroudley Improved Engine Green livery, and sported the luxury of a steam brake. It was soon realised however that when running on a wet day a fully enclosed cab would be beneficial and *Sir Robert* was suitably modified.

(Photo: Jim Haylock)

By the turn of the century significant boiler work was required on both *Susan Jane* and *Sir Robert*, and whilst it was decided to renew the firebox on *Susan Jane's* marine boiler, *Sir Robert* received a significantly more extensive rebuild courtesy of Crowhurst Engineering from Hythe in Kent.

The rebuild included the fitting of a locomotive type boiler, a new fully enclosed cab with rear coal bunker, and to give more leg room to the driver conversion to an 0-4-4T. The loco was re-painted into Great Eastern lined blue livery. A further re-paint has followed but the loco can still be found running at the Wayside Light Railway, these days usually double heading with *Susan Jane*.

Lawrence Martin pictured driving Sir Robert. (Photo: Paul Ash)

GURKHA

Another traditional Tinkerbell design from the first batch, *Gurkha* was finished in a dark green livery and, like *Susan Jane*, with a weatherboard to protect the driver. *Gurkha* was built for Dr Brian Rogers to run on the private Porters Hill Railway in Worcestershire. The line was built on fields next to the family's farm, and featured a main station with covered overall roof, turntable and a long out and back run using a balloon loop. The railway has since been extended to feature another loop through the farm yard giving a very decent length of run.

Gurkha is named in honour of Brian's military connections, as serving in the Royal Corp of Signals regiment during the Second World War Brian saw active service in both India and in the Burma campaign.

Brian was a doctor and is famous in miniature railway circles as the founder of the 7¼" Gauge Society in 1973, initially with a mere 35 members. The society is still running today with over 1,000 members, and *Gurkha* can still be found running on the Porters Hill Railway.

Left: *Brian at the controls of Gurkha.* (Photo: Paul Ash)

Gurkha is pictured here being driven by Brian's son, Seamus. (Photo: Paul Ash)

ADAM

Adam was built to the same specification as *Gurkha*, the main visual differences being a marginally lighter shade of green livery with yellow lining as opposed to the red lining on *Gurkha*.

15

Adam has also seen a much wider sphere of operation and is the most widely travelled of the first five production Tinkerbells. In fact, it is the only one not still to be under the ownership of their original purchaser or their family almost 40 years later.

Initially bought to run on a miniature railway in North Wales, *Adam* moved to the Beale Light Railway near Reading in Berkshire before moving to Scotland and the Ness Island Railway, where it was renamed Uncle John. After a number of years in Scotland the loco moved back south again, this time in private ownership, to be based alongside *Gurkha* at the Porters Hill Railway. It now carries the name *Adam* once again and is in near original condition.

Adam and *Gurkha* have visited the Moors Valley Railway on a number of occasions, often double heading on service passenger trains, a wonderful sight to see.

Left: Adam *on shed at the Moors Valley Railway.* (Photo: Paul Ash)

Adam *and* Gurkha *move off shed to pick up their train.* (Photo: Paul Ash)

16

DEVELOPING THE TINKERBELL CONCEPT
(Can we have a more powerful loco please?)

With *Tinkerbell* and *Talos* already in Jim Haylock's ownership, thoughts turned to how new more powerful locomotives could be constructed for the Tucktonia miniature railway.

MEDEA

Medea was initially constructed as an enlarged version of *Talos* but with an 0-6-2T wheel arrangement, 8½" driving wheels, D shaped smokebox and a fully enclosed cab. *Medea* retained the concept of a marine boiler and Heywood valve gear. Livery was LB&SCR Umber with the name *Medea* painted in gold lettering on the side tanks. To feed the boiler with water *Medea* was fitted with a steam operated water feed pump along with the more normal live steam injector.

Following completion in 1981 the loco was soon converted into a 2-6-2T by fitting an extra set of pony wheels to the front and extending the front running plate. This gave a better weight distribution with the front wheels helping to guide the loco through the curves in the track, reducing wear on the outer rail and giving a smoother ride. A locomotive type boiler replaced the marine boiler in 1985 which made the loco more suitable for intensive operation at Tucktonia.

Medea *pictured before her latest rebuild.* (Photo: Paul Ash)

Medea has since received a full rebuild including fitting 3½" bore cylinders but retaining her Heywood valve gear and LB&SCR Umber livery with gold lettering.

17

Medea is pictured here on the turntable at the Moors Valley Railway following her major rebuild. The matching water cart tender can clearly be seen behind the loco. (Photo: Paul Ash)

SIR GOSS

The next logical step in the development of the Tinkerbell concept was to produce a tender version.

This concept had originally been developed by Roger Marsh and appeared as an artist's impression in the 1980's Severn Lamb catalogue. Called *Severnbelle*, the design showed the loco retaining Heywood valve gear and weatherboard cab, but with a 6-wheel tender.

Right: (Photo: Severn Lamb Ltd)

18

1981 saw Jim Goss (assisted by Jim Haylock) constructing a 2-4-2 tender version of a Tinkerbell in their workshop at Tucktonia. Named *Sir Goss*, the loco featured a locomotive boiler from new, a D shaped smokebox (like *Medea*) but with a large cab and a bogie tender fitted with a tender cab, which gave it a completely unique appearance.

Right: Sir Goss *at the 1992 7¼" Gauge Society AGM at Beale Park near Reading.* (Photo: Paul Ash)

Sir Goss ran at Tucktonia until the line closed in 1985 before moving to the Moors Valley Railway where it was rebuilt as a 2-4-0, fitted with Walschaerts valve gear, a traditional style smokebox and high level running plates. It was painted in fully lined BR Black livery.

Sir Goss *on shed at the Moors Valley Railway following the rebuild as a 2-4-0 with high level running plates.* (Photo: Paul Ash)

Sir Goss left Moors Valley in the early 1990's where it performed a key role in setting up the Mersham Valley Railway in Surrey, before moving to the Swanley New Barn Railway in Kent. Since then *Sir Goss* has clocked up many miles at Swanley hauling passengers on the line that runs from the car park to the main station and back. Currently *Sir Goss* is undergoing a major rebuild that will include a new boiler and valve gear.

SAPPER

Whilst *Medea* took the Tinkerbell concept and made it larger, when constructed the loco was only marginally more powerful than *Talos* as both locos featured the same size cylinders and boiler pressure of 100psi. As train loads increased in size the hunt was on for a larger and more powerful design.

Clearly not a Tinkerbell, *Sapper* was designed and built by Jim Haylock in conjunction with Roger Marsh. The loco was envisaged as representing the War Department 4-6-0 tank engines however it would feature a tender to carry the driver, coal and water. The loco was named *Sapper* in honour of its military roots. The main significant developments of *Sapper* were the use of 3½″ bore x 4¼″ stroke cylinders and Walschaerts valve gear and this would become standard of future builds as we shall see.

Sapper also featured inside frames as per the loco it was representing, however this caused issues with stability and balancing of the wheelsets. The solution was to add significant extra weight to the inside of the frames, along with large internal balance weights on the wheelsets.

Sapper *departs Kingsmere station.* (Photo: Paul Ash)

Sapper proved its worth right from the start and soon became the mainstay of passenger train haulage at Tucktonia. It has since received a significant rebuild including a new chassis, however it can still be seen today running regularly at the Moors Valley Railway hauling full 13 coach passenger trains on a sunny summers day.

AELFRED

The next development of the Tinkerbell concept was completed in 1984 and was the final locomotive to be constructed at the Tucktonia workshops by Jim Haylock. *Aelfred* is a 2-6-4T wheel arrangement and features the larger cylinders and Walschaerts valve gear (as per *Sapper*), a locomotive boiler (as per *Sir Goss*), D shaped smokebox (as per *Medea*) and is complete with larger side tanks, fully enclosed cab with extra legroom for the driver and a rear coal bunker.

The design was inspired by the Vale of Rheidol tank engines and was painted in a fully lined green livery. *Aelfred* represents the full development of the enlarged Tinkerbell concept.

Aelfred *is pictured here looking very patriotic during Her Majesty the Queen's Golden Jubilee celebrations in 2012.* (Photo: Paul Ash)

HARTFIELD

Hartfield was built in 1999 by Martin Colbourne in the workshops of the Moors Valley Railway and represents the updated development of the narrow gauge tank engine design pioneered by Jim Haylock with *Aelfred*.

Hartfield is a 2-6-4T with the same cylinder size and valve gear as *Aelfred* but now with 8½″ driving wheels and the boiler pressure increased to 130psi. Tractive effort is 667 lbs which is three times that of the original Tinkerbell concept on which it is based. *Hartfield* continues to be a regular performer at the Moors Valley Railway being the engine of choice for winter running.

Hartfield *about to depart with another passenger train on the Moors Valley Railway.* (Photo: Paul Ash)

Left: Jason *is a 2-4-4T version*
(Photos: Paul Ash)

Right: Robert Snooks *is an 0-4-4T*

The development of the Tinkerbell concept is pictured here in a single line up. Above L to R: Tinkerbell, Bob, Talos, Medea, Perseus, Robert Snooks *and* Sir Goss. (Photo: Paul Ash)

The same line up viewed from the rear shows the varying cab styles. (Photo: Paul Ash)

23

LEW

Lew is based on the Manning Wardle 2-6-2T loco designed for the Lynton and Barnstaple Railway in Devon and built in 1925. After closure of the line in 1935 *Lew* was used to help dismantle the line before being sold for export to Brazil.

The 7¼" gauge replica of *Lew* seen here was built by David Moore and had the distinction of running on a short line laid on the actual track bed of the Lynton and Barnstaple Railway. During its time at the line the loco used to spend the winters in a small engine shed which did the loco no favours so when it was eventually sold and moved to a new location it was decided to complete a full rebuild and repaint.

Lew is pictured here running on the Wayside Light Railway in Kent. Judging by the camera's and smiles on the faces of the passengers everyone is having a good time. (Photo: Paul Ash)

THE HISTORY OF THE TINKERBELL STEAM LOCOMOTIVE – PART 2
(A potted history lesson in four parts)

At the same time as the construction of the first five production Tinkerbell class locomotives, Roger Marsh also started to sell the drawings and castings under the name of Minimum Gauge Railways, and subsequently from 1978 onwards as Roger Marsh & Co Ltd. This allowed the private builder or enthusiast to chance their hand (and ability) and have a go at building one of these large and powerful locos.

An early adaption on the standard Tinkerbell concept was an inside framed version named *Jacqueline*, built by Mike Sharp and dating from 1979. In the standard Tinkerbell design the main frames of the loco sit outside the wheels which is said to give more stability. *Jacqueline* however has the main frames inside the wheels which gives a completely different look to the loco. Externally the appearance is based on a French Decauville design with the cab supported on stanchions.

Jacqueline *is pictured double heading with* Tinkerbell *at the Moors Valley Railway.* (Photo: Paul Ash)

A strange feature of this loco is the reverser mechanism, which has been set in such a way that you pull the lever back to move forwards and visa-versa, very confusing. Jacqueline is based on a private railway in the south of England.

Right: *Close-up of the unique reverser arrangement.* (Photo: Paul Ash)

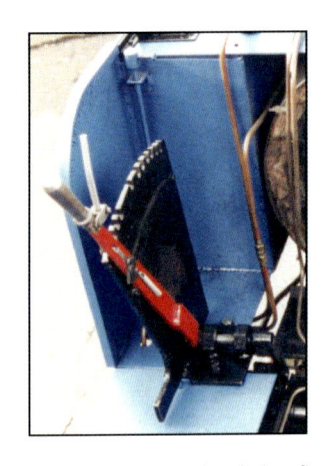

Some of the new Tinkerbell builders aimed to closely replicate the look of the first production locos such as the half cab style first seen on Sir Robert in 1978. *Bratton* was built by David Moore in 1982 and operated for a time on the 7¼″ gauge railway laid on the old track bed of the Lynton & Barnstaple Railway in Devon.

Notable features include a revised twin slide bar and crosshead arrangement, and the tall whistle carried on top of the dome. *Bratton* also has an alternative enclosed cab for those inclement weather days out on Dartmoor.

Bratton *is pictured after overhaul in 2018 by Jim Haylock. The loco is now fitted with the locomotive type boiler formally fitted to* Medea. (Photo: Paul Ash)

Also dating from the 1980's is *Bob* (P Howard 1981) and *Princess* (Edgerton/Horsfield 1989). These locos aimed to closely replicate the look of the original *Tinkerbell* with a weatherboard style cab.

Bob *was built in 1981 and is based at the Hollycombe Museum in Hampshire. The loco is unofficially named after a former driver at the line.* (Photo: Paul Ash)

Princess *spent many years at the Brookside Miniature Railway near Manchester before moving just down the road to the Dragon Miniature Railway.* (Photo: Dom Greenop)

27

Other builders decided to take the basic concept of the sit in narrow gauge tank engine and make a few changes. Notable amongst the newly constructed Tinkerbell's from the 1980's is *Lady Pauline* built by Alwyn Kay in 1982, which added a fully enclosed cab with rear coal bunker onto the existing design. The high cab and drooped cab windows give the loco a sad appearance which is quite distinctive and very unique. *Alwyn's* design has resulted in a large imposing locomotive, great for winter running keeping the driver warm and dry but perhaps a little too toasty in the summer.

Lady Pauline created quite a stir in 1982 when demonstrated at the 7¼" Gauge Society AGM in Runcorn, alongside another early Tinkerbell named *Valerie*, and of course the original *Tinkerbell* being driven by her owner Jim Haylock.

Lady Pauline started life running at the Manor Park Railway in Glossop before moving to the Brookside Miniature Railway where the loco saw extensive use. Since 1999 *Lady Pauline* has been based at the Dragon Miniature Railway near Manchester however at the time of writing she is out of service and due an extensive overhaul.

Lady Pauline *sits quietly between turns at the Dragon Miniature Railway.* (Photo: Dom Greenop)

What soon became apparent very early on was the flexibility of the Tinkerbell design and with the ingenuity of the builder the design could easily be adapted into any number of styles as we have seen over the years. It is this variety that ensures almost every Tinkerbell locomotive is unique.

Some early builders however decided one Tinkerbell loco was just not enough. *Douglas* and *Betty* are a pair of Tinkerbell locos built in 1987 by Stephen Baker to run on the Bow Hills Railway and they demonstrate the concept of an enclosed cab to perfection. This, coupled with a cut away side tank design taken from a similar loco named *Valerie* built a few years earlier, turns them into beautiful locomotives and perfectly suited to the role they perform.

Right: Valerie *pictured on shed at the Silloth Miniature Railway in Cumbria.* (Photo: John Nicholson)

They are also unique in the bodywork being powder coated in blue including the cab roof which might not work for everyone but gives them a trademark in Tinkerbell circles. The pair are still running together today and can often be seen travelling side by side in their own purpose-built trailer when visiting other railways.

Travelling companions Douglas *and* Betty *being carefully unloaded before taking part in a Moors Valley Railway Tinkerbell Rally.* (Photo: Paul Ash)

29

Both the original *Tinkerbell* and the first five production locos featured a marine boiler which once mastered works well for most situations. As we have seen earlier Jim Haylock had modified *Talos* at an early stage by fitting a locomotive firebox to the marine boiler. One of the first applications of the locomotive type boiler on a new build Tinkerbell in the 1980's was *Sian* which was built to run at the Conwy Railway Museum in North Wales.

This Tinkerbell features a D shaped smokebox and fully enclosed cab giving it the now familiar Tinkerbell look. In place of the marine boiler however it was fitted with a locomotive type boiler which made life easier for the driver and ensured free steaming. The line at Conwy is ¾ of a mile in length and operates an intensive service in the summer taking visitors around the site and museum so a strong reliable loco such as *Sian* was just what was required for the line.

Sian has now moved to Germany where the loco has been fully repainted in a striking blue livery and fitted with a fully working steam operated pump for use on air braked coaching stock.

This photo was taken in 2008 on a visit to the UK and shows Sian *in her current condition.* (Photo: Paul Ash)

Into the 1990's and one Tinkerbell builder more than any other stands out as worth a mention and that is Jeff Stubbs from Horbury near Wakefield. Jeff had a passion for building steam locomotives and before he retired from locomotive building in his late 70's had amassed over 20 locomotives to his name. Within that impressive number there was an equally impressive number of 8 Tinkerbell class locomotives. These were built between 1991 (*Lucy*) and 2005 (*Ross*). You got the feeling from talking to Jeff that he loved the hobby and gained real pleasure from his loco building. Sadly, Jeff passed away in 2008.

Jeff Stubbs driving Alice *built in 1995 (not a Tinkerbell) at his beloved Thornes Park in 2005.* (Photo: Paul Ash)

Jeff's philosophy of locomotive construction was simple. He would build the locos he enjoyed operating and would run them at the Thornes Park Railway in Wakefield, before selling them on to fund the next project. A Jeff Stubbs Tinkerbell is very distinctive in its design and construction with their trademark wide radius cab corners, simple pony truck design and usually a beautiful polished dome cover. Over the winter time Jeff fitted a unique feature to his locomotives, namely a steam heated radiator to keep his back warm on a cold day.

Over the years Jeff's Tinkerbell's have travelled the length and breadth of the UK as we shall now see when we look at a few more of his locos in detail.

Lucy was Jeff's first Tinkerbell built in 1991 and is a standard 0-4-2T with a marine boiler and closed cab. *Lucy* spent time at the Hemsworth Water Park in Yorkshire, before moving to the Mortocombe Railway in Berkshire and being renamed Meconopsis.

Lucy *now runs as Meconopsis, named after the Himalayan Blue Poppy. The loco was in the process of being cleaned hence the tin of Brasso on the roof.* (Photo: Paul Ash)

Owd Rosie, built in 1992 and named after a close friend of Jeff's, is something of a Tinkerbell variant and is a 2-6-2T. The loco is extended with a front pony and an extra set of driving wheels, fitted with an extended marine boiler, longer side tanks, larger cylinders and a higher cab giving it a somewhat strange look.

Unfortunately, the long fixed wheelbase has restricted the locos ability to negotiate sharp curves somewhat. When originally constructed the loco was painted in a bright red livery with yellow lining, however it has since received a full rebuild with modified side tanks and a new livery of LB&SCR Umber. *Owd Rosie* runs today on the Swanley New Barn Railway in Kent.

32

Owd Rosie *basking in the sunlight at Swanley ready for her turn on the Santa Specials.* (Photo: Paul Ash)

By the late 1990's Jeff had created a clear formula for his Tinkerbell class locomotives. They were large and powerful, demonstrated his own distinctive design features and were very much suited to the needs of passenger hauling on a busy miniature railway.

A pair of Jeff Stubbs Tinkerbell's built in 1997 and 1999 respectively that portray all the best features of Jeff's locomotives are *Petunia* and *Douval*. *Petunia* demonstrates all the characteristics of a Jeff Stubbs Tinkerbell including the large polished dome cover and large radius cab corners. The loco spent time operating at the miniature railway in Betws-Y-Coed in North Wales before moving south into new private ownership.

Douval, which was formally known as *Lynn Mhari*, started life running at Thornes Park. The loco features full width buffer beams, central heating cab radiator, a fully enclosed cab and has subsequently been fitted with a sliding cab roof hatch for those sunny days. The loco has been a regular performer at the Echills Wood Railway in Warwickshire.

Petunia *is seen here on a visit to the Moors Valley Railway in Dorset.* (Photo: Paul Ash)

Douval *complete with polished dome cover, also pictured during a visit to the Moors Valley Railway in Dorset.* (Photo: Paul Ash)

34

Molly (originally named *Miriam*) was built in 2000 and when constructed featured a locomotive type boiler, green livery, polished dome cover and could be found running of course at the Thornes Park miniature railway. *Molly* is unusual in so far as the spectacle plate is narrower nearer the top, has large round windows and more unusually features side plates sweeping back towards the driver. All of this is best explained by looking at the photo.

With the polished dome cover shining, Molly *is seen after the move south from Scotland but still bearing the SLR initials on the cab.* (Photo: Mike Palmer – Station Road Steam)

After leaving Thornes Park, freshly renamed *Molly* and sporting a new red livery, she moved all the way north to take up residence at the Sanday Light Railway (SLR) on the Isle of Sanday, Orkney, in Scotland. This railway took the honour of the UK's most northerly passenger carrying railway and featured two circuits round the owner's house and tea rooms, before taking a long trip down the side of the driveway to a second station. Whilst at Sanday *Molly* operated alongside another Jeff Stubbs Tinkerbell, a 2-4-2T named *Victoria*. Sadly, the line closed in 2006 amid legal action which is such a shame for a fantastic miniature railway in a wonderful location.

After closure *Molly* and *Victoria* returned south to be sold by Station Road Steam in Lincolnshire. *Molly* however was not sold immediately and ended up in the owners back garden 7¼˝ gauge railway for a number of years before being eventually being sold when the owner re-laid the line to 15˝ gauge.

By 2014 *Molly* had been purchased and moved to the Weston Park miniature railway in Shropshire. *Molly* was then subject to a major rebuild that included a full repaint, boiler re-tube and most noticeably a new half cab. Looking lovely in her new paint scheme *Molly* appears like a new engine, recognisable only by the name and the polished dome cover. Unfortunately, this also demonstrates some of the challenges in compiling accurate historical information where there is very little published material available.

Left: Molly *is pictured here running in her current guise.* (Photo: Paul Ash)

Right: Idris *adopts the same styling as* Molly. (Photo: Paul Ash)

By the late 1990's Tinkerbell class locomotives were growing in popularity and numbers with the Moors Valley Railway (MVR) now holding regular Tinkerbell Gala events every October. This was not only a chance for the resident locos, *Tinkerbell* and *Talos* to stretch their legs on some proper passenger trains, but also the opportunity to see other visiting Tinkerbells running that may more usually be found on private lines and away from public view. Visitors during that time included *Douglas* and *Betty*, *Bob*, *Douval* and *Jacqueline* mentioned previously, along with *Susan Jane, Adam* and *Gurkha* from the original batch.

Other Tinkerbell oddities also appeared such as *Hope* built by Brown/Wade in 1990, which featured a unique design swept back (streamlined) spectacle plate, dumb buffers and a painted name on the side tanks.

Left: Hope *is pictured resting during one of the MVR Tinkerbell rallies.* (Photo: Paul Ash)

Another Tinkerbell class loco to feature at the galas and worthy of a mention is *Ivor* built by Crowhurst Engineering of Hythe in Kent. Built in 1997 *Ivor* features a steel locomotive boiler unusually fitted with copper tubes, 3″ bore cylinders, a spacious cab to accommodate a taller than average driver, and that must have accessory, a steam brake. Ivor manages to portray the classic lines of a Tinkerbell in every respect and now sports a fully lined Great Eastern railway livery complete with white cab roof.

Ivor *is privately owned but resides alongside the other Tinkerbells at the Moors Valley Railway.* (Photo: Paul Ash)

37

Captain Hook is one of a pair of Tinkerbells completed around the turn of the century by David Vere, the other loco being named *Septimus*. They both feature a high seating position for the driver and subsequent high cab and chimney, certainly making it a different experience to drive when compared to a standard Tinkerbell. With the front coal bunker however it gives the loco a well-balanced and pleasing look. *Captain Hook* can usually be found running at the Abeydale Miniature Railway in Sheffield. 2018 saw *Hook*, as the loco is more commonly known, fitted with a new locomotive boiler pressed to 150psi and replacing the previous marine style.

Left: *Painted in lined black livery,* Septimus *stands on the turntable at the Abeydale Miniature Railway.* (Photo: Ben Harris)

Captain Hook *is seen here on a short freight train on the Moors Valley Railway in Dorset.* (Photo: Paul Ash)

38

1998 saw the 'Tinkerbell 30th Anniversary Gala' at the Moors Valley Railway to celebrate 30 years since the building of the original *Tinkerbell* steam locomotive. The event was attended by visiting Tinkerbell class locomotives *Gurkha* and *Susan Jane* from the original batch of five, resident locos *Talos*, *Ivor* and of course *Tinkerbell* herself, along with a visit from her builder, Roger Marsh. Also in steam was enlarged Tinkerbell derivative 2-6-4T *Aelfred* and the 2-4-0 + 0-4-2 Garrett named *William Rufus*. Honour of driving *Tinkerbell* was given to Martin Colbourne and Russ Green. A visiting diesel was also in attendance.

MVR 'Tinkerbell 30' line up L to R – Tinkerbell, Talos, Gurkha, Susan Jane *and* Ivor. (Photo: Paul Ash)

Unfortunately, *Susan Jane* suffered a broken fire bar during the day and had to return to shed to have the fire dropped and the bars repaired, an easy task in the well-equipped workshop. She soon returned to service.

Left: Susan Jane *suffered a failure and is seen having her fire dropped.* (Photo: Paul Ash)

DRIVING & FIRING TINKERBELL
(Peter Jackson shares his experiences)

My first meeting with *Tinkerbell* was in 1997, when for the grand summer gala for that year *Tinkerbell* hadn't been assigned a driver. Surprisingly, I was then allocated to be her driver for the weekend. *Tinkerbell* is certainly a different beast to the other Moors Valley Railway (MVR) engines.

Fist job of the day – a cup a tea. Peter is seen having a quick break before starting a hard day's work driving Tinkerbell. *(Photo: Peter Jackson)*

Tinkerbell, although named after Roger Marsh's dog, has many of the characteristics of the fairy from Peter Pan. *Tinkerbell* is very definitely a slightly naughty engine who likes to remind her drivers that she definitely requires your entire attention.

So how do I drive *Tinkerbell*?

Let's start with the Fire…

To fire *Tinkerbell*, you must forget what you think you know about firing a locomotive. With a marine style firebox, *Tinkerbell* has a unique method of firing when compared to the other engines. You need 3 tools. 1 – Shovel, 2 – Poker and 3 a solid boot.

To fire *Tinkerbell* the best thing to do is regularly clear out the ash from the ash pan using the poker, then ensure the fire is nice and flat (again using the poker) before you put a fresh round of coal on. Now using the shovel put the coal on the fire. Making sure you fire the front (sometimes you will need the poker for a third time to push the coal further forward). Once you have filled the fire with coal, shut the door. If the door shuts, you have not put enough coal on the fire. Using the shovel (or hands) add more coal to the fire until door doesn't quite shut. Now push it shut hard, on occasions this might mean the use of tool 3!!!

You have now fired *Tinkerbell*...

The magical Blower!

Blower use (the way of creating a draft to draw the fire when then engine is not chuffing) on most MVR engines is rare, and just like them *Tinkerbell* doesn't usually need the blower. *Tinkerbell* however does like the refreshing breeze of steam up her chimney at all times, so its sensible to keep her happy and have the blower cracked open. Failure to have the blower cracked is likely to upset *Tinkerbell* and mean suddenly she just will not steam that well, usually when you have a full load of passengers on the train!!!

You should now have a nice hot fire and plenty of steam.

Using Peter's firing technique, this is what you are looking for.

Left: Tinkerbell's *safety valve lifts whilst waiting at a red signal just outside Kingsmere station.* (Photo: Russ Green)

Injectors – yes please

With the fire working nicely, you will be using water and need to inject water into the boiler using one or both of her injectors. Unlike the other MVR Engines, *Tinkerbell* is fitted with an inbuilt "too much water" turning off setting (made by her ability that if her water level is higher than her clack valves she will stop injecting water). *Tinkerbell* at all times likes to keep you on your toes and make sure you are paying full attention. So, she ensures that at random times through the day one of the injectors will just stop working for a while. Just as suddenly she will allow it to work again.

Now it's time to chuff

Once mastering the fire, blower and injectors you can start chuffing. With *Tinkerbell* it is important to oil her VERY regularly, in particular her valve gear. Even with this regularly oiling at points during the day *Tinkerbell* will talk to you, changing her gently clanging noise from the valve gear to a completely different note. This is *Tinkerbell* saying she suddenly wants more oil. It is important to give her more oil as soon as possible, otherwise for no explanation you can be sure that your pressure will drop, and she will not be a nice engine to drive.

While driving, *Tinkerbell* will tell you which cut off to use on the reverser lever. Whereas one trip this might be notch 3, on another, this might be notch 2. Each time she will inform you of what notch she is happy with by her wiggle while running, never ever put her in a notch that makes her judder, she will tell you off!!!

Tinkerbell loves to talk to her driver in any way she can. A trip where steam is being produced in abundance and the water level is high is her way of saying she is happy, whilst a trip where pressure is low is her way of saying she wants a bit more of your attention. She can change her view at any point, so good trip can be followed by a bad trip before a good trip again.

Way back in 1997 I was told by Jim (her owner) that if you have a bad trip, one of the things you must do, is polish an inch. By this you find a dirty bit on her paint (and bits will be dirty during the day) and polish it. Again, this is somehow important to *Tinkerbell*… No one knows why!

Tinkerbell *double heading with* Sapper. (Photo: Josh Walker)

MEMORABLE DAYS WITH TINKERBELL

(More experiences by Peter Jackson)

Since 1997, I have been one of the two main drivers of *Tinkerbell*, and every day on *Tinkerbell* is always exciting and a highlight (even if she is in a mood!). There are far too many to write about, so I have selected a few.

Tinkerbell and an 8 carriage train in the summer of 2003

Firstly, it is worth noting *Tinkerbell* cannot easily pull a full summer MVR set of 11 coaches, a luggage van and guards van. During the summer of 2003, management noted that the sale of *Tinkerbell* memorabilia (in particular spoons and badges) had for some reason dried up. As *Tinkerbell* is not suited for a big train, and we run 3 big trains in the summer, it's not ideal to have *Tinkerbell* running on her own in the summer. However, by the grace of British weather, the next day weather forecast was showing rain until 1pm and then cloudy thereafter. I thus asked if I could get her out as "train 2" in the hope of selling some memorabilia as it was unlikely we would need 3 full length trains (and probably wouldn't actually need 2).

As predicted the rain was heavy in the morning and *Tinkerbell* was steamed up. At 10:30 she was coupled to her 8 car train, and ready to go. At 10:31 the rain stopped, and by opening at 10:45 the sun was out and in glorious form. Suddenly the quiet day was going to be a lot busier than expected. *Tinkerbell* did her best, with full trains all day (with people deliberately waiting for the little yellow one!).

At 4pm as the passenger numbers started to decline, *Tinkerbell* decided she had done enough, and she decided to not make steam and came to a stand on the approach to Kings tunnel. She did however seem to know that management had set the route for her to come off that trip and go to shed. So maybe it was her way of complaining!!! At the end of the day, a spoon and several badges had been sold, along with a fair few of "TINKERBELL SAVES THE DAY" books.

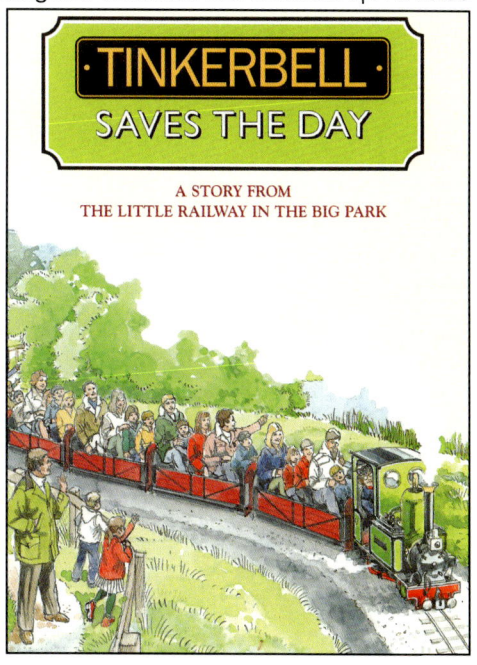

Tinkerbell and *Talos* go for a big train in 2006

After the "success" of the adventure in 2003, I was lucky enough to persuade management to have her out the following year, but this time double heading with Jason, this lead to the following year a day double heading with Horton, and then we arrived at 2006.

The then summer driver (Andy) was up for trying a full service train with the two small ones. It is very important to note, that to do a full 11 coach, a luggage and guards van, neither engine could actually pull all that on its own all day, especially if the coaches are full.

Off we went running the first train of the day and proved they could do it. As the day wore on *Talos* was having a few fits, but *Tinkerbell* was steaming well. At 1pm management decided it was probably worth having a spare "large" steam engine on standby in case one of the engines failed. At this point people were taking bets when we would need to be replaced by the larger engine. Interestingly most people were going for around 2:30pm. This was obviously overheard by the two engines, as 2:30pm came and went, and even with *Talos* not steaming at her best we were trundling round quite happily.

At 4pm the "spare" engine was sent back to shed, and then we managed to take the last train of the day. To my knowledge it was the first time (at least since 1997), that *Tinkerbell* and *Talos* has run a full set from first train to last train. Certainly, both myself and the other driver were shattered but also with huge smiles.

On board shot of Tinkerbell & Talos *double heading.* (Photo: Russ Green)

THE HISTORY OF THE TINKERBELL STEAM LOCOMOTIVE – PART 3

(A potted history lesson in four parts)

Into the new millennium and Tinkerbell locomotives proved to be as popular as ever with over 50 Tinkerbell or variants being recorded as in existence at that time. Roger Marsh, however, had now stopped selling drawings and castings by this time and Chris Fincken had taken over marketing them as part of Fincken Miniature Railways. For the record a full set of 18 drawings would set you back £80 including P&P, whilst a full set of castings would be a massive £845.21.

Built in 2002 by Joe Nemeth for the Oldown Miniature Railway near Bristol, *James* is a traditional looking 0-4-2T Tinkerbell differing from the original batch only in so far as it has been fitted with a locomotive type boiler from new. Operating out of the Oldown Country Park, the miniature railway featured a main station with turntable and engine shed, and a run of approximately ½ mile through the park with a balloon loop through the woods before returning to the main station. The line was open for 3 years before closing when the owner decided to move location and set up a new line in 10¼" gauge at the Cattle Country Park in Gloucestershire.

James *stands on the turntable at the Oldown Miniature Railway after the last train of the day.* (Photo: Paul Ash)

An identical sister locomotive to *James* was started by Joe at the same time but never finished and was sold on to Station Road Steam in 2004. Work completed included main frames, wheels, a complete boiler (from Bell Boilers) and a full set of castings.

A full set of castings. (Photo: Station Road Steam)

The loco then spent a time with a buyer in Scotland, before returning to Joe in 2006 via Station Road Steam, still in a part built condition. It was purchased in 2007 by the author who completed the build over the next 8 years with the loco steaming in 2015. It is named Hestia and is based at the Moors Valley Railway in Dorset where it can often be seen double heading with *Talos* on a passenger service or running on special event days. It is hoped one day to be able to have the two (brother and sister) Tinkerbells re-united and double heading on a passenger service.

Looking sparkling Hestia *was only a few weeks old when this photo was taken in 2015.* (Photo: Paul Ash)

Like Jeff Stubbs before him, Eric Walker of Kirton in Lincolnshire was another prolific Tinkerbell builder. Eric had led an interesting life including working as a signalman on the railway, travelling extensively and lately repairing furniture among other things. He did not start building steam locmotives until the age of 70, starting with another Roger Marsh design, Romulus. Amazingly he partitioned off part of the front room for the workshop, installed a 1938 Colchester lathe, and proceeded to lay the Kirton Light Railway (KLR) round the garden using homemade concrete sleepers and steel rail.

Eric built his first Tinkerbell, loco No 2 in 2002. The loco was fitted with a marine boiler built by Sam Ward, cylinders from Eric's own pattern but otherwise it was unmistakenly a Tinkerbell. Eric used the loco on his garden railway which featured tight curves and steep gradients so it was certainly a test of the new loco. Eric sold the loco in 2004 to Mike Palmer at Station Road Steam from where it went to the Bath and West Railway in Somerset.

During this time it was totally rebuilt by the new owner, extended (no more toasted knees) and converted into a 2-4-2T. New valve gear was also designed and fitted. Today the loco is named *Sara* and is owned by Recreation Railways based in Somerset, where it can be seen running at a number of different events in the local area giving rides to children (or anyone else who buys a ticket).

Sara, *following her intensive rebuild, sits outside the shed at the Bath & West Railway.* (Photo: Paul Ash)

Eric certainly did not stop his Tinkerbell building after that, on the contrary he became much more ambitious. 2003 saw completion of a Tinkerbell based *Garrett* (No 3), followed in 2004 by a Fairlie derivative (No 4). This featured an articulated front power bogie, with the remaining loco being pure Tinkerbell. This however highlighted a shortfall of the design as the loco seriously lacked weight over the front drivers which limited adhesion. On leaving Kirton the loco spent a brief time at the Lackham Woodland Railway before being purchased by Recreation Railways.

Tinkerbell Fairlie KLR No 4 outside the shed at the Lackham Woodland Railway. (Photo: Paul Ash)

By now Eric was in his mid 70's but he continued to produce locomotives at a rate of one every 6 months or so. KLR No 8 was a covered cab Tinkerbell, whilst 12, 13 and 14 were Eric's standard Tinkerbell, all completed and sold via Station Road Steam in 2012. These later Tinkerbell's (KLR 12-15) are very much garden railway machines with smaller cylinders, side tanks and less substantial valve gear.

Nonetheless they have proved to be good workers if a little agricultural in their engineering and lacking finesse in their finishing. KLR No 15 was the last Tinkerbell built by Eric after he severed a finger on one of the machines. Sadly, Eric passed away in 2014 aged 87 having completed an amazing nine Tinkerbells in 13 years.

Rebranded and renumbered, ex KLR No12 stands outside the shed on a visit to the Moors Valley Railway. (Photo: Paul Ash)

Peter has been rebuilt from KLR No 8 with the chassis extended by 4 inches, new side tanks, dome cover, new pipework and expertly repainted. Peter clearly demonstrates how fine an Eric Walker Tinkerbell can look. (Photo: Dave Bayliss)

49

Not one to be constrained by the original Tinkerbell design, and a lover of the Isle of Mann narrow gauge steam locomotives, Paul Frank built a pair of locomotives named *Douglas* and *Dragonfly* in 2003 as 2-4-0T and closely resembling the Isle of Mann outline. They feature marine boilers, a longer fixed wheelbase as the rear driver needs to be under the driver's seat, high up inclined cylinders, polished dome covers and of course that unmistakable rounded cab roof.

In service the locos have proved to be good reliable runners however do suffer a little from the high cylinder position inducing some rocking motion, especially when under load.

Douglas can still be found operating at the Conwy Valley Railway Museum at Betws-Y-Coed in North Wales, whilst *Dragonfly* has now moved to the Dragon Miniature Railway near Manchester.

Right: Dragonfly *shows off the unique square cab windows.* (Photo: Paul Ash)

Standing outside the shed at Conwy, Douglas *has totally captured the look of an Isle of Mann narrow gauge steam locomotive.* (Photo: Paul Ash)

50

Merlin is a Tinkerbell from the Narogauge stable of Jim Haylock and was built in 1993. It has the traditional 0-4-2T wheel arrangement, 2½" cylinders, 9¾" diameter driving wheels, marine boiler, Heywood valve gear and sports a steam operated water feed pump.

The main difference from a standard Tinkerbell is that it has lengthened frames (+ 3 inches) and a rear coal bunker arrangement. Originally named *Merlin*, the loco was re-branded whilst in private ownership to *Steamy Dream*, before being sold on again to operate at the short lived Bolebrooke Castle miniature railway.

Since the closure of that line it has moved back into private ownership and has been renamed as *Merlin* once again.

Branded as Steamy Dream, Merlin *rests quietly in the yard.* (Photo: Paul Ash)

Perseus started life as a concept back in 1998 and was envisged to be an exact copy of *Merlin* with the same size wheels, cylinders etc, the only difference being a locomotive boiler from day one. As the project developed and after long discussions with Jim Haylock the wheel size was reduced to 8½", cylinder size increased to 3½", and Walschaerts valve gear fitted in place of the more normal Heywood type. Changes were also made to the design of the smokebox and cab windows including fitting a 2" thick smokebox saddle and 1" thick front buffer beam to balance the weight of the driver and coal bunker. The only part of the new locomotive therfore that remained a direct copy of *Merlin* was the cab and tanks as these were cut from the exact same pattern.

The result was a significantly more powerful loco that was able to handle the full 13 coach trains that run on the Moors Valley Railway, whilst maintaining the looks of the original Merlin. *Perseus* was completed in 2006 and is currently undergoing it's 10 year overhaul. It is expected to return at the end of 2018.

Perseus *outside the shed at the Moors Valley Railway ready for a days work.* (Photo: Paul Ash)

Perseus *even has the honour of having it's potrait painted by Mike Haylock. A very good likeness.* (Painting: Mike Haylock)

52

Another Tinkerbell locomotive builder starting to establish a reputation is Peter Beevers. Peter only started building Tinkerbell locomotives in 2008 and already has six to his name. Peter has taken a slight twist on the proven formula with all his 0-4-2T Tinkerbells featuring larger cylinders, a rear coal bunker and revised slide bar arrangement. This has given them the nickname of a Beeverbell.

First up we have *Zebedee* built in 2012. This is clearly based on the original Tinkerbell design but with the larger cylinders giving a more rugged appearance. The small weatherboard would obviously give no protection to the driver, so it is unsurprising to know that since moving to the Ness Island Railway near Inverness in Scotland, *Zebedee* has now gained a larger half cab. With this came a new paint scheme (British Railways black number 68194) and a new name, *Chrissy*.

Zebedee *is pictured shortly after construction in 2012.* (Photo: Paul Ash)

Two more of Peter's Tinkerbells next, this time from 2012 and 2015 respectively. Both this time feature a covered cab with either rear stanchions or a closed back, along with the usual rear coal bunker.

Zanna was built for a private customer at the same time as *Zebedee*, whilst *Joan* dates from 2015 and operates at the Barnards Miniature Railway in Essex. This railway is relatively new however is quickly developing into something quite special with a run of over 1 mile out and back from the main station called Burtonshaw (with covered overall roof) to Angel Green and back. The line runs through lovely parkland, features steep gradients and turntables at the main stations. *Joan* is joined by a number of other locos on the line including two Exmoor built 0-4-2T.

53

Peter van Zeller drivng Phil Ashworth's Zanna *at the Kyre Valley Railway.* (Photo: David Nicholson)

Joan *ready to depart with a train load of passengers from Angel Green on the Barnards Miniature Railway.* (Photo: Paul Ash)

Narogauge, under the leadership of Jim Haylock has been building Tinkerbell and variant locomotives since the early 1980's and over that time turned out countless variations on a theme. 2013 however saw Tim Woron taking over as the lead role for these Tinkerbell builds, both for private customers, and based closely on that of *Talos*, already operating at the Moors Valley Railway.

Tertius started life as a part built Tinkerbell loco bought by a private customer with the contract awarded to Narogauge to complete the build. The loco when delivered to the workshops of the Moors Valley Railway was little more than a rolling chassis and can be distinguished from *Frederick* in so far as *Tertius* retains a set of buffers from the original construction. *Frederick* was a completely new construction but both locomotives were progressed through the works at the same time and share many common parts.

Both locos feature all the improvements already applied to *Talos* over the years such as a locomotive boiler, ball raced valve gear and extended frames. *Tertius* has been built with a covered cab and rear stanchions however this can easily be converted to run without the cab roof on a sunny summers day. *Frederick* on the other hand was built with the front spectacle plate only but can now be seen running cab-less as in the photo.

Construction took place in the workshop at Moors Valley. Tertius *can be seen on the right whilst* Frederick *is on the left.* (Photo: Paul Ash)

Tertius *outside the carraige shed at the Moors Valley Railway ready for the next tain service.* (Photo: Paul Ash)

Frederick *rests between duties on the Barnards Miniature Railway.* (Photo: Paul Ash)

BUILDING A TINKERBELL
(Some notes on construction)

As already mentioned in Part 2 of the history of the Tinkerbell steam locomotive, drawings and castings have been made available since the late 1970's for the private locomotive builder to construct their own Tinkerbell class locomotive, with many of these already completed and running on railways throughout the UK and abroad. Instrumental in the development of these designs has been Jim Haylock at the Moors Valley Railway. This is the story of the construction of Tinkerbell locomotive *Hestia*, built by Paul Ash with the help and assistance (inspiration) of Jim Haylock. *Hestia* is close to the original *Tinkerbell* design with 2½" bore cylinders, weatherboard cab and the now standard locomotive boiler.

Construction starts with the two main frames made from ½" thick steel plate. These days CAD design and water jet cutting makes this a very easy task to produce a set of frames which can then be tack welded together and drilled before being separated for assembly. The standard Tinkerbell chassis gives an overall length of 64" however for the taller driver this can easily be extended by adding 3" into the foot well. Both *Tinkerbell* and *Talos* have had this modification in later years although it is easier to complete before the frames are cut. In order to balance the weight of the driver sitting nearer the back of the loco additional weights are added just behind the front buffer beam.

The standard design of axle box on a Tinkerbell features large ball race bearings sitting on machined axle boxes and running with large horn guides. A pair of springs sit above each axle box. Whilst this works fine for the standard design when you add additional weight from the locomotive boiler and a larger cab the lack of adjustability can cause issues. Jim Haylock has developed a system of hangers to support the axle boxes which uses 4 springs per axle box as opposed to the 2 on a standard Tinkerbell. The rear pony truck is pivoted off the cab floor with sprung plungers pressing on a cross stretcher on the chassis. On *Hestia* an inside framed rear pony truck is fitted, once again using fully ball raced bearings. Alternatively an outside framed pony truck can be fitted and you will see a fair mix of both styles being used on other Tinkerbell locomotives depending on the preference of the builder.

Tinkerbell was built with 10" wheel size whist this was increased to 10¾" for the production batch. *Hestia* utilises 9¾" diameter driving wheels with outside fly cranks. The wheels are cut from steel blanks machined to profile with water jet cutting again used for the fly cranks before machining. They are assembled with a tight interference fit. Assembly of the rolling chassis takes place next with the fitting of the wheels and coupling rods. The chassis is bolted to the front and rear buffer beams, also ½" thick steel plate, with a cross stretcher in the middle of the chassis.

Left *Basic chassis, wheelsets and boiler components for* Hestia. (Photo: Paul Ash)

Additional stretchers support the firebox on the boiler, with a full width stretcher supporting the rear of the cab floor. The final stretcher supports the rear pony truck plungers. The chassis is painted at this point before fitting the wheel sets. The rolling chassis is complete.

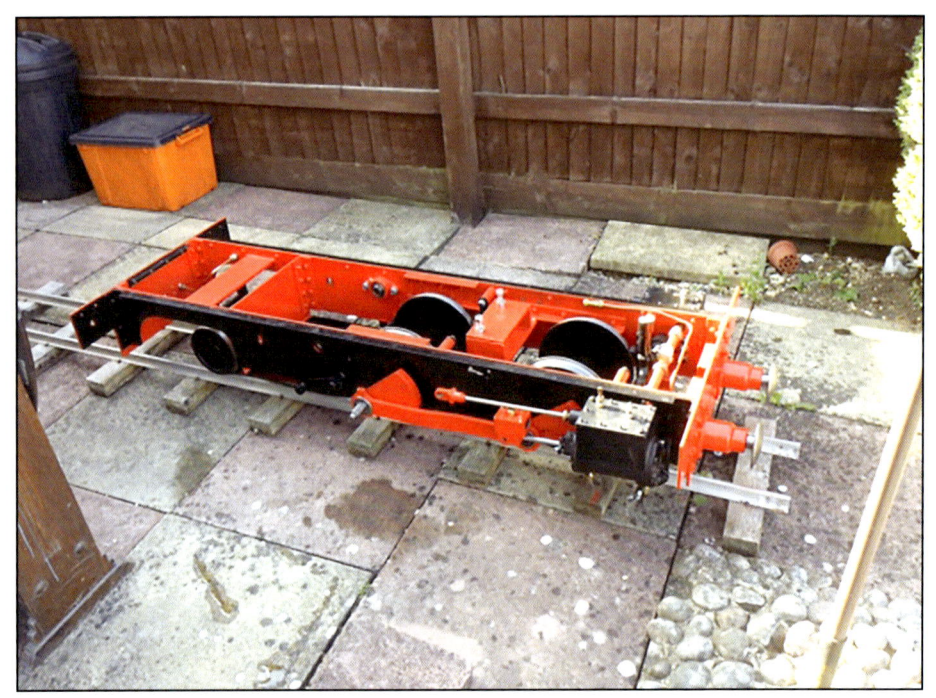

The completed and painted rolling chassis. (Photo: Paul Ash)

Attention now turns to the cylinders. *Hestia* features 2½″ bore x 4¼″ stroke and these are machined from cast iron and incorporating the valve chests as a single casting. The pistons are made from cast iron and fitted with a pair of steel Clupet piston rings to provide the seal in the bores. The slide valves are machined from bronze and fitted to the valve rods allowing the valves to float in the steam chest whilst still providing a steam tight seal under steam pressure.

Valve gear on *Hestia* is the version of Heywood gear fitted to the first five production Tinkerbells. This features a die block connected to the reverser and linked through banana links to the rear of the connecting rod. Here is where modern technology is one again used with bronze bushings being replaced by roller bearings running of hardened steel pins. The die block also uses a pair of ball race bearings to ensure smooth operation and reduce wear. Lubrication is provided by oil nipples on all moving joints and the result is a valve gear that is without any slop in the joints whilst remaining free running.

Assuming all the calculations and measurements have been correct then by now there will be a completed chassis running on air. All being well the chassis will run on 10psi with the aim to be 4 even beats from the exhaust.

Next consideration is given to the boiler, smokebox and ashpan. On the locomotive boiler fitted to *Hestia* there are 30 tubes of ¾" diameter. The boiler is set to 100psi working pressure however this involves a hydraulic test to 150psi before fitting to the loco. The ashpan sits between the loco wheels and directly under the grate. There are as many different ashpan designs as there are Tinkerbells however the main focus is to ensure sufficient primary air to the fire whilst keeping the hot ash from falling onto the track. *Hestia* features a damper on the ashpan to restrict the primary air flow to the fire and keep the fire quiet when standing.

The smokebox sits at the front of the boiler to collect the ash that has passed through the tubes but not managed to make it all the way up the chimney. A flat plate spark arrester is fitted in the smokebox. The chimney on *Hestia* is finished off with a cast brass top. The smokebox door is traditional Tinkerbell with a central brass knob and four removable clamps. The boiler rests on the chassis stretchers at the rear and the smokebox at the front and is secured by clamps that allow fore and aft movement only. The smokebox on *Hestia* is made from 3mm thick steel plate rolled into a cylinder and welded to the front of the smokebox. Additional 3mm plating is then welded to the floor of the smokebox to provide extra strength and longevity from the corrosion caused by the ash drawn through from the fire.

If you're following the story so far then stick with it as we get to the exciting bits, the body work and cab. *Hestia* features two full length running plates and the tanks and cab then sit on these. All the body work is made from 3mm steel plate welded construction and once again CAD drawing and laser cutting make this a quick and simple task. *Hestia* once again follows the traditional design with both side tanks holding water but the right hand tank also holding the coal supply. Internally the tanks are sealed with fibreglass and liquid rubber sealant. A weatherboard cab is fitted. The boiler wrapper is made from 1.5mm steel plate and is fitted over the boiler lagging, whilst a fibreglass dome cover provides the finishing touch.

Next up is painting which in the case of *Hestia* is LMS crimson lake applied by spraying over the usual combination of shot blast steel, filler, primer, undercoat and top coat. Copious amounts of sanding is required to gain a good finish. Lining is yellow with black border.

Left: Hestia *pictured prior to painting.* (Photo: Paul Ash)

Final tasks before the first steaming is the pipework. Use is made of commercial fittings here including the water gauges, pressure gauge, safety valves and all cab controls. The copper pipework and connectors are also commercially available. Finally comes another hydraulic test to ensure no leaks and then the steam test. With any new boiler it can take 10 steaming's or so to clear all the dirt from the boiler and avoid priming particularly when put under load. *Hestia* was first steamed in 2015.

Hestia *with her builder, Paul Ash, in 2015 on the day of the first steaming of the loco.* (Photo: Paul Ash)

THE HISTORY OF THE TINKERBELL STEAM LOCOMOTIVE – PART 4

(A potted history lesson in four parts)

Whilst *Tinkerbell* is clearly a very British narrow gauge steam locomotive, it is worth noting that Tinkerbell class steam locomotives are not just confined to the UK. There is in fact significant contingent operating out of the Stoomgroep West Zuiderpark near the Hague in the Netherlands.

Left to Right – *No 1* Tiny Tim, *No 2* Blaagje, *No 3* Brutus *and No 4* Cor (Photo: Wilfred Buijs)

Construction came about, like so many new build Tinkerbell projects, following a visit by Cor Cammenga to the UK and the Moors Valley Railway. All four Tinkerbell locos feature similar looks with half cabs, distinctive dome covers and the tapered high chimney.

Built over a period of years from the early 1990's the most recent of these is named *Cor*, after it's builder Cor Cammenga. *Cor* features a very unique steam operated bell alongside the more usual whistle.

Right: *Steam operated bell fitted to* Cor. (Photo: Wilfred Buijs)

Another group of Tinkerbell's worth a mention are a trio built by Frank Stephen to run on the Royden Park Miniature Railway at Wirral, Merseyside. The loco's are named *Peter George*, *Let's Rumble* and *Tinker* and all were based on the Roger Marsh original, but with much added input following a visit made to the Moors Valley Railway.

They are all constructed as 0-4-4T to give extra legroom in the cab, and all feature a unique cab design that is tapered in at the top. *Tinker* and *Peter George* can still be found operating at Royden Park, however *Let's Rumble* has now moved to a new owner in Northern Ireland.

Peter George *pictured at the Royden Park Miniature Railway.* (Photo: Frank Stephen)

The design has proved popular as a similar loco was built by Peter Lapworth in 2012, named *Paxton*. Uniquely it is fitted with an electric water feed pump powered by a battery hidden in the rear coal bunker.

Right: Paxton (Photo: Dave Bayliss)

There is another group of 0-4-2T narrow gauge steam locomotives that are worthy of inclusion in this story although not strictly classed as Tinkerbells. These are the locomotives built by the Exmoor Steam Centre at Bratton Flemming in Devon.

Although perhaps more famous for building locos in gauges up to 15", the design has been scaled down to 7¼" gauge to create a very sturdy locomotive designed for commercial use on a regular basis. Exmoor have turned out 13 of these since 1999 with the majority being 0-4-2T and fitted with a half cab.

Amy Louise (Left) *and* Jean (Right) *are standard Exmoor locos. Both have previously operated on the Brookside Miniature Railway.* (Photos: Paul Ash)

Jeremy *on the other hand features a closed cab and a somewhat German narrow gauge outline. Underneath it remains 100% Exmoor.* (Photo: Paul Ash)

Another Exmoor loco with a different appearance is *Jools* from the Beer Heights Light Railway, operating out of Pecorama in Devon. Formally a standard Exmoor named *Samatispur*, the loco has recently undergone a major overhaul in the railways own workshops and was re-named by none other than the singer Jools Holland himself.

Jools *rests outside the shed on the Beer Heights Light Railway in the company of* Dickie, *which was built by David Curwen in 1976.* (Photo: Callum Darraugh)

The cab of an Exmoor engine features large chunky controls and makes good use of commercial fittings.

Right: *Here we see the cab of* Hunton, *a covered cab Exmoor based at the Wayside Light Railway.* (Photo: Paul Ash)

2008 saw the 'Tinkerbell 40th Anniversary Gala' take place at the Moors Valley Railway in Dorset, the home of *Tinkerbell* herself. The weekend was very special indeed as it bought together the first five production Tinkerbell class locomotives alongside the original *Tinkerbell* for the first time since their construction in 1978.

Advantage was taken of a photo opportunity to line up all six engines side by side in front of the main coach shed, along with a group sounding of whistles. Each loco also carried a commemorative headboard for the duration of the event.

Line up Left to Right: Susan Jane, Sir Robert, Adam, Gurkha, Talos *and* Tinkerbell *herself.* (Photo: Paul Ash)

Right: *Roger Marsh at the controls of* Tinkerbell *at the 40th Anniversary Gala in 2008.* (Photo: Paul Ash)

This brings the story of the history of Tinkerbell steam locomotives to an end with one of the newest. *Mrs. Darling* was built is 2017 by Peter Jackson and painted in a Cadbury purple livery. The loco shares many commonalities with the original *Tinkerbell* built almost 50 years earlier.

Completed in 2017 Mrs. Darling *is seen on shed at the Moors Valley Railway.* (Photo: Paul Ash)

At the time of writing research has shown there at least another three new locomotives under construction and due for completion in the very near future. Some of these will display the more traditional style however one new locomotive is due to take on a colonial style including a wooden cab roof which should be interesting. Planning is also well underway for the 50th Anniversary celebrations due to take place on the 22nd and 23rd September 2018 at the home of *Tinkerbell*, the Moors Valley Railway in Dorset.

I wonder if Roger Marsh could have foreseen way back in 1968 when *Tinkerbell* was first steamed that 50 years later we would be celebrating such an event, or that over 100 different Tinkerbell class locomotives have been constructed and are operating in as far away as Australia, or even that *Tinkerbell* would still be in steam on a regular basis.

Here's to the next 50 years!

PHOTO INDEX *(page numbers)*

Adam	8, 16, 65	Lady Pauline	28
Aelfred	21	Lew	24
Alice	30	Meconopsis	32
Amy Louise	63	Medea	17, 18, 23
Betty	29	Merlin	51
Blaagje	59	Molly	35, 36
Bob	23, 27	Mrs Darling	66
Bratton	26	Owd Rosie	33
Brutus	61	Paxton	62
Captain Hook	38	Perseus	23, 52
Cor	61	Peter	49
Dickie	64	Peter George	62
Douglas	29	Petunia	34
Douglas (2)	50	Princess	27
Douval	34	Robert Snooks	22, 23
Dragonfly	50	Sapper	20, 42
Frederick	55, 56	Sara	47
Gurkha	8, 15, 16, 39, 65	Septimus	38
Hartfield	22	Severnbelle	18
Hestia	46, 58, 60	Sian	30
Hope	37	Sir Goss	19, 23
Hunton	64	Sir Robert	8, 13, 14, 65
Idris	36	Susan Jane	8, 12, 39, 65
Ivor	37, 39	Talos	9, 10, 11, 23, 39, 44, 65
Jacqueline	25, 26	Tertius	55, 56
James	45	Tinkerbell	4, 5, 6, 7, 9, 23, 25, 39, 40, 41, 42, 44, 65
Jason	22	Tiny Tim	61
Jean	63	Ursula	3
Jeremy	63	Valerie	29
Joan	54	Wren	2
Jools	64	Zanna	54
KLR No 12	49	Zebedee	53
KLR No 4	48		

BIBLIOGRAPHY

Barratt, J. & Shaw, L., 2007-2017. *Steam Locomotives - Dragon Miniature Railway.* [Online]
Available at: http://www.freewebs.com/dragonrailway/
[Accessed 14 July 2018].

Bellchamber, D. H. M., n.d. *Tinkerbell Saves the Day.* s.l.:Moors Valley Railway.

Fincken, C., 1999. *Fincken Miniature Railways.* s.l.:C. Fincken & Co Ltd.

Holroyde, D. & Munday, N., 2010. *Miniature Steam Railway Locomotives in the British Isles.* s.l.:LCGB.

Ltd, S. L., n.d. *Severn Lamb Ltd - Model Makers to the World.* s.l.:John Towersey Design Assoc. Ltd.

Marsh, R., 2013. Tinkerbell - the Early Days. *Engineering in Minaiture,* 34(12), pp. 418-419.

Museum, C. V. R., n.d. *Locomotives & Rolling Stock.* [Online]
Available at: http://www.conwyrailwaymuseum.co.uk/
[Accessed 14 July 2018].

MVR, n.d. *Locomotives.* [Online]
Available at: www.moorsvalleyrailway.co.uk
[Accessed 17 July 2018].

Smithers, M., 1995. *Sir Arthur Heywood and the Fifteen Inch Gauge Railway.* s.l.:Plateway Press.

Smithers, M., 1995. The Moors Valley Railway. *Engineering in Miniature,* 16(8), pp. 228-231.

Smithers, M., 1995. The Moos Valley Railway. *Engineering in Minaiture,* 16(9), pp. 281-283.

Walker, E., 2014. *My Story Volume 5.* s.l.:KLR.

Whisstock, L. P., 1993. New "Tinkerbells" in Holland. *7 ¼" Gauge News, Issue 67,* pp. 29-30.

White, R. F., 1991. *Cromar White Ltd - Miniature Railway Engineers.* s.l.:Cromar White.

With special thanks to the following people: David Nicholson, Jim Haylock, Andrew Webb, Tim Woron, Andy Jefford, Russ Green, Peter Jackson, Frank Stephen, Dave Baylis, Wilfred Buijs, Dom Greenop, Ben Harris, Mike Palmer, Peter Beevers & Callum Darraugh, without which this book would not have been completed.